# THE HUMANOID AND ROBOTICS REVOLUTION

## A GUIDE TO THE $7 TRILLION FUTURE

JIM MELLON

Fruitful
PUBLICATIONS

This edition was first published in November 2025



Registered Office: Fruitful Publications, 26 Prentice Street, Lavenham, Sudbury, England, CO10 9RD.



Material contained in this book is for general information purposes only. It is not provided as professional advice and readers should consult a professional where appropriate. This book is not intended to be relied upon by individual readers in making (or refraining from making) any specific investment decisions. The price of shares and investments and the income derived from them can go down as well as up, and investors may not get back the amount they invested. Past performance cannot be relied upon as a guide to future performance. The author and Fruitful Publications do not accept any liability for any losses suffered by any user as a result of any such decision.

A catalogue record for this book is available at the British Library.

ISBN 978-1-7392505-6-0

Fruitful
PUBLICATIONS

# Acknowledgements

We would like to extend our thanks to the following for all the help, support, and information they provided during the production of this book:

AI – specifically **Claude, Manus and Chat GPT** – over the past few months I have become very good friends with AI – it would have been wrong for me not to use it for help with this book – and I did! It's fickle but improving with age!

Most importantly – **Trish Wilson** for her indefatigable and indispensable work on this. It wouldn't have happened without her.

**Philip Boigner** – for his very diligent and valued advice.

**Sir James Bucknall** – I appreciate his insightful input on the key military aspects of robot tech and other helpful editing.

**Denham Eke** – For offering, as always, pithy and appropriate guidance.

**Bob MacDonald** – For excellent commentary – much appreciated my fellow Orielenses!

**Mark Peberdy** – Thank you Mark – very helpful.

**Nicki Poole** – Whilst everyone else was busy with this book, Nicki kept the show on the road!

**James Sibley** – Who picks up new themes with ease and has been super helpful.

And of course.

**Laura Walker** for her important and invaluable input.

"Robots will be a big part of our future, and the benefits they bring – in productivity, safety and even companionship – are only just the beginning"

– *Bill Gates*

# Table of Contents

# Overview

The global robotics sector stands at a transformative juncture. Humanoid robotics is emerging as a genuinely revolutionary technology. It is poised to fundamentally reshape global manufacturing, healthcare, logistics, defence, agriculture, and numerous other industries.

This brief analysis hopefully presents a compelling investment case for the robotics industry, with particular emphasis on humanoid robotics, which Citigroup forecasts will evolve into a $7 trillion global market by 2050[1]. This represents approximately 6%[2] of the current worldwide GDP - which is huge - bigger than food and energy. The robotics industry is poised to become one of the largest global economic sectors within the next two or so decades.

I am convinced that robotics is the single best way to play the AI boom - not data centres, not chips, not power sources - but robots as an augmentation or substitution for many tasks undertaken currently by humans. Mobile, self-charging, multi-tasking, cheap and efficient, non-unionised and uncomplaining - there will literally be billions of them around the world within 20 years[1].

The convergence of artificial intelligence, advanced materials science, precision manufacturing, and declining component and battery costs has established optimal conditions for widespread robotics commercialisation. What was previously confined to science fiction is rapidly materialising into industrial reality, with companies such as Boston Dynamics, Tesla, and Unitree Robotics advancing rapidly to establish industry leadership across real-world applications spanning from military operations to elderly care facilities.

Recent developments throughout 2025 have dramatically accelerated adoption timelines for robotics across multiple sectors. The 2025 World Robot Conference in Beijing featured over 600 humanoid robots, representing a sevenfold increase from the previous year. Meanwhile, Chinese manufacturers have achieved breakthrough pricing with humanoid robots now available from Unitree Robotics for under USD $5,900. This represents a cost reduction exceeding 90% compared to previous generation systems, fundamentally altering the economic equation for robotics adoption across industries where they were previously considered economically unviable.

According to the International Federation of Robotics, global robot installations reached a record 4.3 million units in 2023, representing a 10% increase from the previous year despite global economic uncertainties[3]. Recent data from August 2025 indicates that major Chinese manufacturers industrial robot exports grew nearly 60% to $746m in the first half of 2025 driven by manufacturing shifts to Vietnam and elsewhere[4].

Humanoid robotics represent among the fastest-growing segments across the multiple application domains of the robotics market. Unitree Robotics, a Chinese humanoid company, reported in July 2025 that orders for humanoid robots have already grown by more than 200% year-on-year[4]. The expected compound growth rate for humanoids from 2023 to 2028 is expected to hit 53%, with a further 42% compound growth to 2033· There has never been an industry like it. It will easily surpass AI[5].

The economic implications are profound. Business case payback periods for robotics have compressed dramatically from the historical 5 to 7-year timelines, with some applications now achieving payback in well under 18 months[1]. This compression is driven by improved robot capabilities, reduced integration costs, and the urgent need to address labour shortages in key industries creating economies of scale. For example, a leased robot can cost c.$300 per month. Think of it like this; $300 for a worker who operates 24/7, never takes a holiday except to be maintained, probably by a fellow robot, and never calls in sick - now, that's a bargain!

After decades of incremental progress, robotics technology has reached commercial viability with remarkable speed across diverse application domains. Manufacturing task success rates have reached 80-99%[6] for specific applications, including pick and place, sorting, assembly, and inspection. Defence applications used in controlled environments, agricultural implementations and elderly care applications show improved effectiveness and high levels of reliability.

The integration of Vision-Language-Action (VLA) models (AI) with reinforcement learning has become the mainstream approach, enabling robots to perform sophisticated tasks across many applications. However, industry experts note that Google's new AI video generation model, Veo, introduced in May 2025, which creates and incorporates audio and image prompting, presents an interesting discussion point among robotics companies regarding whether its underlying AI architecture could be even more optimal than current VLA model-combined with reinforcement learning. In practice, hybrid systems are emerging - combining VLA's semantic flexibility with general-purpose, interactive robots, and Veo's more task-specific control precision.

## Investment Landscape and Market Opportunities

The investment landscape encompasses publicly traded robotics leaders, high-growth private companies, and the extensive supply chain ecosystem that supports robotics manufacturing. Currently, robotics ETFs manage over $3 billion in combined assets, which is really pretty small, with the top 30 robotics companies having a combined market capitalisation of $2.1 trillion, although that figure is distorted by the inclusion of Tesla *(market cap approx. $1.12tn)*, which is absolutely not a pure robotics company!

The geographic distribution shows the United States represents 45% of companies and 65% of robotics market capitalisation, with strengths in AI technology, software development, and venture capital. Meanwhile, Asia-Pacific markets, particularly China, lead in manufacturing and cost-effective production. As mentioned above, the US figures are probably somewhat overstated since robots are not yet significant revenue contributors for many large technology companies included in robotics indices.

McKinsey's latest robotics research reveals that advances in embodied AI, foundation models, and behavioural cloning are enabling the transition from specialised robots to general-purpose systems capable of completing diverse, unrelated tasks across different settings. This represents a fundamental shift from the traditional single-task automation paradigm to flexible, adaptable robotic systems. The most significant commercial opportunities lie in specialised applications where humanoid robots can leverage their general-purpose architecture for high-value tasks across defence, agriculture, security, care of the elderly, and transportation sectors[7].

**Foundation AI** models *(large-scale machine learning models trained on vast datasets that serve as general-purpose engines for a wide range of downstream tasks)* represent the most critical technology driving general-purpose robotics capabilities. This means that, after being trained on vast amounts of data, these models can identify patterns that allow robots to perform multiple tasks with varying degrees of optimisation across different applications. The integration of multimodal foundation models enables robots to perform actions based on both visual inputs and language inputs, such as spoken commands, at least partly addressing what researchers call "Moravec's paradox"- the observation that computers excel at high-level cognitive tasks but struggle with physical tasks that humans find simple.

## Chinese Market Dominance and Supply Chain Control

China has emerged as the dominant force in humanoid robotics manufacturing, with domestically produced component ratios reaching approximately 80%. This vertical integration has enabled cost advantages that are driving global adoption and creating significant opportunities for investors who understand the supply chain landscape. The Chinese government has implemented unprecedented policy support for robotics adoption, including consumer purchase subsidies of 1,500 RMB (c. US $210) per unit and corporate purchase subsidies of 250,000 RMB (c. US $35,000) per unit[8].

Major Chinese humanoid robot companies, including Astribot, Booster Robotics, Galbot, Kepler, LimX, Paxini, RobotEra, Spirit.ai, Unitree Robotics, and UBTech, have demonstrated rapid product evolution compared to just a year ago, with many new products launched around the 2025 Robot Conference. The strong family and consumer visitor traffic observed at the event suggests that near-term educational, companion, and entertainment demand potential could be significant, indicating that if robot products are useful, affordable, and safe enough, consumer demand potential could be very substantial, aside from the other areas of demand.

## Sector-Specific Applications and Growth Projections

**Medical Robotics** represents a $23 billion global market share in 2024 and an expected 15% compound annual growth rate (CAGR) between 2025-2034[21], driven by surgical robotics, rehabilitation devices, and assistive technologies. The benefits of medical robotics are compelling: increased precision with lower invasiveness, higher levels of consistency and lower risk of human fatigue-related errors, and enhanced recovery and patient outcomes. The integration of humanoid capabilities into medical robotics is creating new opportunities to address the global themes of ageing populations and shortages in healthcare workers, with recent demonstrations of humanoid caregiver robots capable of monitoring and gently repositioning immobile patients.

**Elderly Care Applications** address the global demographic challenge, with robots demonstrating exceptional utility in mobility assistance, health monitoring, and companionship – including haptic and artificial skin applications that provide a more human feel and enhance trustworthiness. The economic impact extends beyond labour cost savings to encompass improved quality of care and enhanced independence for elderly populations.

**Manufacturing and Industrial Sectors** remain dominant in robotics, with steady growth in traditional systems now being challenged by more adaptable humanoid robots. Collaborative robots (Cobots), which enable multi-robot teamwork, now account for over half of global installations, reflecting a shift toward system-wide efficiency. Industrial robots drive faster production, better quality, and safer workplaces. Industrial robots drive faster production, better quality, and safer workplaces with 'Dark Factories' now becoming more widespread globally, where robots work 24/7-without the requirement for light, food, or sleep, whilst maintaining consistent speed, accuracy, and predictability.

**Agricultural Transformation** will showcase humanoid robots performing tasks such as planting, tending, and harvesting with precision that often exceeds human capabilities. These systems can withstand harsh environmental conditions and perform detailed, plant-level interventions that were previously impossible with traditional mechanised farming equipment.

**Security and Defence** remain strategically vital areas for humanoid robotics, offering enhanced force multiplication, reduced risk, and improved operational efficiency. The U.S. Department of Defense anticipates significant cost savings from autonomous systems, driven by lower recruitment, training, and casualty-related expenses. These robots can operate longer in hazardous environments, reducing the need for human presence. For example, while a soldier in Afghanistan cost around $850,000 annually, a tactical robot may cost just $230,000 - highlighting the potential for long-term savings and reduced battlefield casualties[9].

## Outlook and Strategic Implications

Industry experts now predict that the "ChatGPT moment" for robotics, when robots can complete tasks with generalised skillsets in completely new environments, is 2 to 3 years away, significantly closer than recent previous estimates of 5 to10 years. This acceleration is driven by advances in AI model architecture and the increasing deployment of robots for collecting real-world data.

The investment implications are profound. Early-stage investors who position themselves correctly in this emerging ecosystem stand to benefit from what could be the largest technological disruption since the advent of the internet. However, navigating this landscape requires a sophisticated understanding of technological capabilities, supply chain dynamics, regulatory frameworks, and geopolitical considerations that will shape the industry's evolution.

CHAPTER 1:

# Market Analysis and Growth Projections

## The Global Robotics Market Explosion

According to the International Federation of Robotics, global robot installations reached a record 4.3 million units in 2023, representing a 10% increase from the previous year despite global economic uncertainties. However, these figures represent only the beginning of what industry experts predict will be an exponential growth phase as humanoid robots transition from experimental prototypes to commercial reality.

Interest in robotics technology has increased five- to six-fold over the past 12 to 18 months[10]. As evidenced by company reports and press publications, this is reflected in conversations with investors. Historically, robotics were confined to certain repetitive use cases, with concerns about return on investment, implementation costs, and the complexity of integrating them into operations. Robots were once literally caged in factories; no longer, they are mobile and increasingly being designed to perform far, far, more than repetitive, simple tasks.

The transformation from confined, single-purpose industrial robots to mobile, multi-functional systems represents a paradigm shift that is opening entirely new market segments. Modern robots can navigate complex environments, interact safely with humans, and adapt to changing requirements without extensive reprogramming. This flexibility has enabled deployment in previously inaccessible sectors such as hospitality, retail, healthcare, and domestic services.

Market research indicates that the robotics industry will experience a CAGR of approximately 15-20% from 2025 to 2033[19] across most segments. Humanoid robotics are showing even higher growth rates of 40-50% annually[11]. This growth is being driven by several key factors: declining hardware costs, improved AI capabilities, labour shortages in key industries, and increasing acceptance of robotic solutions among consumers and businesses.

The economic drivers behind this growth are compelling. Labour shortages in developed countries, particularly in manufacturing, healthcare, and service industries, are creating urgent demand for robotic solutions. Simultaneously, the ageing population in many developed nations is driving demand for elderly care robots, while the need for increased productivity and efficiency is pushing adoption in manufacturing and logistics sectors.

## Humanoid Robotics: The $7 Trillion Opportunity

To reiterate, Citigroup's latest comprehensive analysis projects that the global humanoid robotics market could reach $7 trillion in annual revenue by 2050, representing approximately 6% of current global GDP[1] and establishing robotics as one of the largest industries on the planet. This projection has been corroborated and potentially even understated by other recent analyst views[12].

Robot adoption will continue to increase as technological progress in key areas such as AI and dexterity expands usage, and price reductions lead to faster payback periods. Citigroup forecasts the number of robotic units will show 13% CAGR between 2025 and 2035, led by growth in Service Robots, Autonomous Vehicles and Industrial Robots; from 2035 to 2050 similar growth rates in robotic units worldwide will be led by a 48% CAGR in humanoids as they become integrated across all sectors of the economy, with household penetration increasing globally, and with the emergence of robot-as-a-service (RaaS) business models.

The humanoid robotics market, projected at $7 trillion annually by 2050, could surpass the global automotive industry and rival the energy sector, reshaping global economic structures. This growth will create new value chains and business models. Companies leading in robotics manufacturing, components, software, services, or integration stand to gain significantly. Geographically, Asia-Pacific - especially China - leads in manufacturing and cost efficiency, while North America and Europe excel in advanced AI and high-value applications.

## Technology Readiness and Commercial Viability

The robotic market opportunity has expanded significantly beyond initial projections, with multiple distinct segments showing accelerated growth and commercial viability. Manufacturing task success rates (in other words, the percentage of output performed purely by robots) have reached 80-99.5% for selective applications, including pick and place, sorting assembly, and inspection.

## Industrial Robotics

Industrial robotics is evolving from fixed systems to flexible, AI-powered Cobots with a $43.8 billion market value and expected 11.5% CAGR between 2024-2033[20]. These mobile, easily reprogrammable machines can safely work alongside humans and integrate into existing workflows. Companies like Universal Robots (a Danish company), with over 100,000 global installations, demonstrate the commercial success of this shift. The adaptability of Cobots is lowering adoption barriers, especially for small and medium-sized manufacturers.

## Service Robotics

Fortune Business Insights stated that the market size for service robots is rapidly scaling and projected to grow to $90.09 billion by 2032 exhibiting a CAGR of 19.2% since 2024 moving from pilot projects to full commercial deployment. In 2025, Galbot deployed 200 wheeled humanoids in Meituan (a food delivery company)'s warehouses - up from zero the year before - highlighting this shift. Service robotics spans cleaning, security, delivery, and customer service, with companies like Amazon and others proving their viability across sectors, especially during the COVID-19 pandemic.

## Humanoid Robotics

Humanoid robotics ($7 trillion potential market by 2050) has reached commercial viability faster than expected, driven by general-purpose AI that enables robots to adapt, learn, and perform diverse tasks without extensive reprogramming. Unlike traditional robots, humanoids can operate in human environments using existing tools and infrastructure, reducing deployment costs. Recent demonstrations across industries - from manufacturing and logistics to entertainment and elderly care - highlight their versatility and transformative potential.

## Geographic Market Dynamics

The global robotics landscape is shifting toward Asia-Pacific, with China emerging as a leader in humanoid robotics manufacturing. This is driven by strong government support, integrated supply chains, and rising market demand. As of 2025, Chinese robots use around 80% domestically produced components, enabling cost-effective production and broader commercial deployment. Strategic policies, including research funding and regulatory support, have fostered a robust innovation ecosystem, positioning China at the forefront of global robotics.

## North America: AI Leadership and High-Value Applications

The U.S. leads in AI and robotics software, supported by a strong venture capital ecosystem and a focus on high-value, specialised applications like medical, defence robotics and AI software platforms. Companies like Symbiotic Inc. have achieved significant market capitalisations through warehouse automation solutions, while PROCEPT Bio Robotics focuses on surgical robotics for urological procedures.

However, it faces cost and scale competition from China. Tesla's Optimus program has scaled back production targets, reflecting broader challenges. With a market valued at $18.7 billion and 422,000 robot installations in 2023, North America prioritises advanced AI integration and premium applications where performance outweighs cost.

## Europe: Precision Engineering and Regulatory Leadership

Germany leads the European market in advanced manufacturing applications, with a strong focus on collaborative robotics and precision engineering. The European approach emphasises safety standards, regulatory compliance, and premium applications rather than cost competition. Companies like KION Group AG focus on material handling and warehouse automation, alongside established brands such as Linde, STILL, and Dematic.

European companies have established leadership in collaborative robotics, with companies like Universal Robots pioneering safe human-robot collaboration. The European market also leads in developing regulatory frameworks for the deployment of robotics, which may become increasingly important as robotics adoption accelerates globally.

## Emerging Markets: Leapfrog Opportunities

Developing economies are rapidly adopting robotics, especially in sectors facing labour shortages and productivity demands. Countries like India, Brazil, and Southeast Asian nations are bypassing older automation technologies in favour of advanced, AI-powered systems. Despite lower labour costs, the need for efficiency and quality is driving robotics adoption in manufacturing, agriculture, and services.

CHAPTER 2:

# Technology Landscape and Innovation Drivers

## Foundation Models and Embodied AI

Advances in multimodal foundation models are enabling robots to understand both visual and language inputs, allowing them to generalise across tasks and environments. This marks a shift from traditional programming (which requires specific instruction for each task) to adaptable, embodied AI that interacts with the physical world. By integrating large language models, robots can now follow natural language commands, ask questions, and explain actions - making them more accessible and reducing the need for specialised programming skills.

## Hardware Advances and Component Integration

The hardware revolution in robotics encompasses advances across multiple domains: sensors, actuators, processors, and materials science. These improvements optimise robot dexterity, sensing capabilities, and power management while simultaneously reducing costs and improving reliability. The convergence of these hardware advances with AI software capabilities has created the conditions for commercially viable general-purpose robotics.

## Advanced Sensing and Perception Systems

Modern robots are equipped with advanced multimodal sensors - such as gyroscopes, cameras, tactile sensors, LiDAR (stands for Light Detection and Ranging – effectively creates distance maps of a scene), and IMUs (Inertial Measurement Units) – to perceive and interact with their environments more intelligently. These systems enable precise movement, object recognition, and delicate manipulation without traditional fixtures. Neuromorphic sensors, inspired by the human brain, can process data efficiently and in real time, while event-based cameras improve responsiveness by only transmitting changes in the visual field. Together, these innovations significantly enhance robotic perception and adaptability[13].

## Actuator and Motion Control Innovations

Advanced actuator systems are key to achieving human-like dexterity in robots. Combining rotary and linear actuators allows for precise, flexible manipulation, with Chinese firms like Sanha Intelligent Controls leading in component production. Modern actuators now include technologies like pneumatic muscles and shape memory alloys, enabling natural, safe movement. High-torque, lightweight actuators are crucial for mobility and efficiency in humanoids. Integrated with advanced control algorithms and machine learning, these systems support complex tasks through adaptive stiffness, smooth motion, and continuous performance improvement.

## Processing Power and Computing Architecture

NVIDIA dominates the robotics computing space with its Jetson platform, widely used for real-time Edge AI deployment. Its full-stack ecosystem supports model training, simulation, and deployment. Modern robots require powerful, low-latency computing to process sensor data and control actuators, making Edge computing essential. GPUs (Graphic Processing Units), TPUs (Tensor Processing Units), and neuromorphic chips are increasingly integrated to handle complex AI tasks efficiently (also at lower power consumption), enabling responsive and intelligent robotic systems.

## Materials Science and Durability

Advanced materials science has enabled the development of robots that can withstand harsh environmental conditions while maintaining precision and reliability. Agricultural robots, for instance, must operate in dusty, wet conditions, exposed to chemicals and extreme temperatures. Carbon fibre composites, advanced alloys, and engineered plastics provide the structural integrity needed for robotic systems while minimising weight. This is particularly important for humanoid robots, where weight directly impacts mobility, battery life, and safety in human interaction.

Hydrogen fuel batteries are starting to be used in drones and logistics robots. These present some technical challenges, including miniaturizing components like fuel cell stacks and power converters, transporting fuels cells to drone launch sites, efficient, and automated refuelling for logistics robots. Despite these, fuel cell adoption is expected to grow across drones, logistics robots, unmanned underwater vehicles (UUVs), and aircraft.

Smart materials that can change their properties in response to environmental conditions are opening new possibilities for adaptive robotics. Shape memory alloys can provide variable stiffness, while electroactive polymers can function as artificial muscles. These materials enable robots to adapt their physical properties to different tasks and environments.

## Behavioural Cloning and Imitation Learning

Robots can now learn by watching humans - either live or in videos - in different contexts. The visual input is incorporated into models that enable robots to replicate human actions, even in complex, multi-step activities.

This approach represents a dramatic departure from traditional robotics programming, which required painstaking step-by-step instructions specifying the movements of each robot actuator in sequence. To understand the complexity reduction, consider that traditional programming for blueberry picking would require instructions for each muscle equivalent in the robot's arm, wrist, hand, and fingers, specifying how far, how fast, and with what force to move, all based on multimodal, real-time feedback from sensors.

Emerging clean food technologies - like precision fermentation and cultivated meat - require constant, precise management of large, pressurised bioreactors. Robots are well-suited for these demanding tasks, enabling cost-effective, around-the-clock production.

Imitation learning allows robots to acquire complex behaviours by observing humans, other robots, or video demonstrations. Recent algorithmic advances let robots extract key features from these demonstrations and generalise them to new situations, significantly reducing the need for manual programming.

## Reinforcement Learning and Adaptive Behaviour

Reinforcement learning enables robots to improve their performance through trial and error, learning from the consequences of their actions. This approach is particularly effective for tasks where optimal strategies are not immediately obvious or where environmental conditions vary significantly. Robots can learn to optimise their behaviour for specific objectives, such as minimising energy consumption, maximising task completion speed, or improving safety.

Multi-agent reinforcement learning enables robots to work together effectively, allowing them to learn how to coordinate their actions and share information. This is particularly important for applications involving multiple robots, such as warehouse automation or search and rescue operations.

## Modular and Reconfigurable Systems

The development of modular robotic systems enables rapid reconfiguration for different tasks and applications. Rather than designing specialised robots for each application, modular systems can be assembled from standardised components to meet specific requirements. This approach reduces costs, improves flexibility, and accelerates deployment timelines.

## Simulation and Digital Twins

Advanced simulation environments have become essential for robotics development, enabling testing and training of robotic systems in virtual environments before deployment in the real world. NVIDIA's Isaac Sim platform, Unity's robotics simulation tools, and other advanced simulation environments provide physics-accurate virtual worlds where robots can be trained and tested safely and efficiently.

## Digital Twin Technology

The creation of virtual replicas of real robotic systems and their environments enables continuous monitoring, optimisation, and predictive maintenance. Digital Twins can be used to test new behaviours, optimise performance, and predict potential failures before they occur in the real world.

The development of photorealistic simulation environments has enabled the training of computer vision systems using synthetic data. This approach can generate unlimited training data with perfect ground truth labels, addressing one of the major bottlenecks in AI development for robotics.

## Cloud Robotics and Distributed Intelligence

Cloud robotics enables robots to offload complex AI processing to cloud servers while maintaining real-time control locally. This hybrid model combines vast cloud computing power with the responsiveness needed for physical tasks. Distributed intelligence allows multiple robots to share data and coordinate actions via the cloud, supporting swarm robotics where large groups of robots collaborate by exchanging sensor data and synchronising movements.

## Edge Computing

Edge technology refers to innovations that enable data processing and analysis to occur closer to the source of data generation, rather than relying solely on centralized cloud or data centre infrastructure. To balance performance and responsiveness, modern systems integrate edge computing with cloud robotics. Time-sensitive tasks are handled locally, while the cloud manages more demanding AI workloads. This approach ensures both advanced capabilities and real-time control, essential for effective robotic operation.

## Development Tools and Frameworks

The development of sophisticated robotics development frameworks has dramatically reduced the complexity and time required to develop robotic applications. The Robot Operating

System (ROS) and its successor ROS 2 provide standardised interfaces and tools for robotics development, enabling developers to focus on application-specific functionality rather than low-level system integration.

Modern robotics development environments offer visual programming interfaces, simulation tools, and debugging capabilities, making robotics development accessible to a broader range of developers. These tools enable rapid prototyping, testing, and deployment of robotic applications.

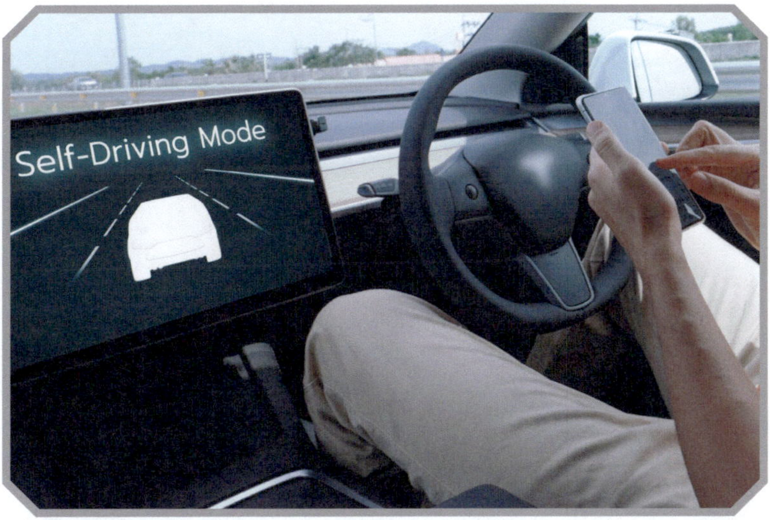

CHAPTER 3:

# Surgical Robotics and Precision Medicine

The medical robotics sector is one of the most promising and rapidly expanding applications of robotic technology, with the global market valued at $23.1 billion and experiencing a CAGR of 15-20%[14] (2024-2030). This growth is driven by the convergence of several factors: an ageing global population, increasing demand for minimally invasive surgical procedures, the need for enhanced precision in medical interventions, and the growing shortage of skilled healthcare professionals in many developed countries.

Surgical robotics has evolved from experimental systems to essential tools that improve patient outcomes across multiple medical specialties. Companies like PROCEPT Bio Robotics Corp focus specifically on surgical robotics for urology procedures, demonstrating the commercial viability of specialised medical robotics applications.

Modern surgical robots provide surgeons with enhanced dexterity, tremor elimination, and magnified visualisation that surpasses human capabilities. The integration of AI-powered systems enables real-time analysis of surgical procedures, providing guidance and alerts that improve safety and outcomes. These systems can process multiple data streams simultaneously, including imaging data, vital signs, and surgical instrument positioning, to provide comprehensive situational awareness.

## Minimally Invasive Surgery Revolution

The shift towards minimally invasive surgical techniques has been accelerated by robotic systems that can perform precise movements through small incisions. This approach reduces patient trauma, shortens recovery times, and decreases the risk of complications compared to traditional open surgery. Robotic surgical systems can filter out hand tremors, scale down surgeon movements for enhanced precision, and provide 3D visualisation of the surgical field with magnification capabilities that exceed human visual acuity.

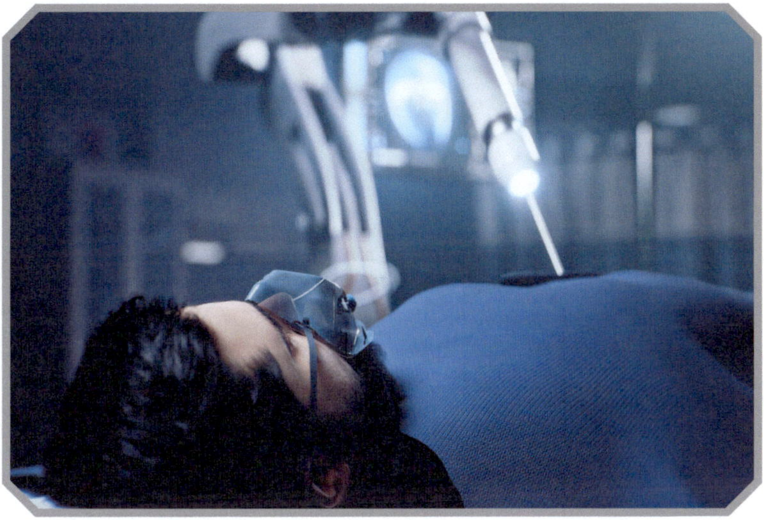

## AI-Enhanced Surgical Planning and Execution

Modern surgical robots incorporate AI systems that can analyse pre-operative imaging data to create detailed surgical plans, identify potential complications, and suggest optimal approaches for specific procedures. During surgery, these systems can provide real-time guidance, monitor vital signs, and alert surgical teams to potential issues before they become critical. The integration of machine learning algorithms enables these systems to improve their performance over time, learning from successful procedures and adapting to different patient anatomies and surgical scenarios.

## Rehabilitation and Assistive Technologies

Rehabilitation robotics represents a rapidly growing segment focused on helping patients recover from injuries, strokes, and other medical conditions. These systems provide consistent, precise therapy that can be customised to individual patient needs and progress, whilst reducing the burden on healthcare professionals and improving patient outcomes.

Robotic rehabilitation systems can provide therapy 24/7, enabling more intensive treatment regimens that accelerate recovery. Assistive robotics technologies are transforming the lives of individuals with disabilities, offering enhanced mobility, improved communication, and increased independence.

Exoskeleton technology represents one of the most promising areas of rehabilitation robotics. Companies like Lifeward focus specifically on exoskeleton technology that can assist patients with mobility impairments. These systems can provide powered assistance for walking, standing, and other mobility functions, enabling patients with spinal cord injuries, strokes, or other neurological conditions to regain independence and improve their quality of life.

Humanoid robots designed for mobility assistance can detect when a patient is at risk of falling and provide immediate stabilisation, whilst also encouraging independent movement and exercise. The integration of AI systems enables these robots to learn individual patient movement patterns and provide personalised assistance that promotes recovery whilst ensuring safety.

Brain-computer interfaces are enabling direct neural control of robotic systems, opening new possibilities for individuals with severe disabilities. Neurological Rehabilitation and Recovery Robotic systems can provide repetitive, precise movements that are essential for neural pathway recovery following stroke, traumatic brain injury, or other neurological conditions. These systems can adjust resistance, speed, and range of motion based on patient progress, providing therapy that adapts to individual recovery patterns. The consistency and precision of robotic therapy systems often exceed what human therapists can provide, whilst detailed data collection enables precise monitoring of patient progress and therapy optimisation.

## Hospital Automation and Patient Care

The integration of robotics into hospital operations extends far beyond surgical applications to encompass patient care, logistics, cleaning, and administrative functions. This comprehensive automation addresses multiple challenges facing modern healthcare systems: staff shortages, infection control, operational efficiency, and patient safety.

Humanoid robots designed for patient care can perform a wide range of tasks that traditionally require human staff, including patient monitoring, medication delivery, mobility assistance, and basic care activities. Demonstrations of a 30kg Humanoid caregiver robots capable of gently repositioning immobile patients highlight the potential for robotics to address physically demanding aspects of patient care whilst maintaining safety and dignity.

Robotic systems can provide continuous patient monitoring, collect vital signs, movement data, and other health metrics without requiring constant human supervision. The integration of AI analysis enables these systems to identify patterns and trends that might not be apparent to human observers, potentially improving early detection of complications or changes in patient status.

Automated medication systems can reduce errors, ensure timely delivery, and maintain detailed records of medication administration. Robotic systems can verify patient identity, confirm medication dosages, and provide reminders for patients who self-administer medications. These systems can also manage medication inventory, track expiration dates, and ensure proper storage conditions for temperature-sensitive medications.

## Infection Control and Cleaning Robotics

Cleaning and disinfection systems have become increasingly important, particularly in the wake of the COVID-19 pandemic. These systems can provide consistent, thorough cleaning and disinfection of patient rooms, operating theatres, and common areas using UV light, chemical disinfectants, or other sterilisation methods. The precision and consistency of robotic cleaning systems can reduce infection rates whilst freeing human staff for direct patient care activities.

## Logistics and Supply Chain Management

Hospital logistics robots can manage the movement of supplies, equipment, and materials throughout healthcare facilities. These systems can transport medications from pharmacies to patient units, deliver meals, move clean linens, and collect waste materials. The integration of these systems with hospital information systems enables real-time tracking of supplies and equipment, reducing waste and ensuring the availability of critical items when needed.

## Market Opportunities and Investment Landscapes

The medical robotics market presents compelling investment opportunities across multiple segments. Surgical robotics companies are achieving significant market capitalisations, while rehabilitation and assistive technology companies are experiencing rapid growth. The ageing global population is driving demand for medical robotics solutions, creating a large and growing market opportunity.

The regulatory landscape for medical robotics is complex, with stringent approval processes like FDA clearance in the U.S. and CE marking in Europe. While these requirements can be time-consuming and costly, they also create high entry barriers that offer competitive advantages to companies that successfully navigate them. Achieving regulatory approval not only validates product safety and effectiveness but also provides market protection and credibility.

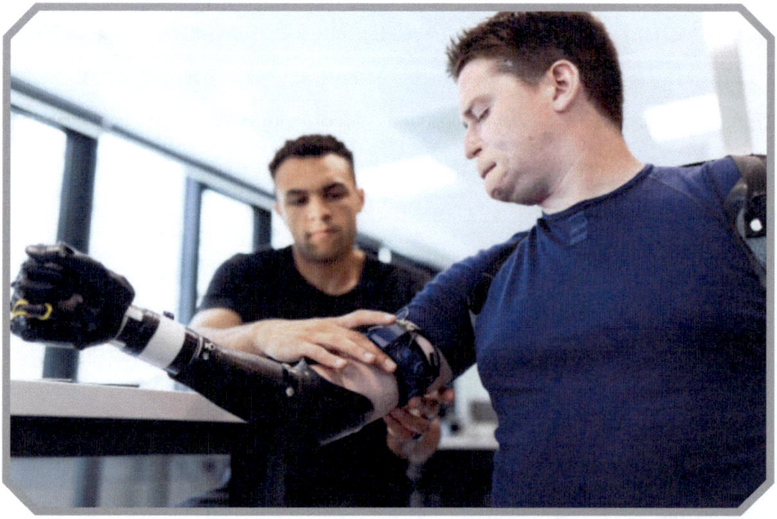

## Investment Trends and Market Dynamics

The medical robotics market exhibits strong geographic concentration in developed countries with advanced healthcare systems and regulatory frameworks. The United States represents the largest market, driven by high healthcare spending, advanced medical facilities, and a supportive regulatory environment for medical innovation. Europe and Japan also represent significant markets, whilst emerging economies are beginning to adopt medical robotics as their healthcare systems develop, and costs decline.

## Future Growth Drivers and Market Expansion

Several factors are expected to drive continued growth in medical robotics, including the ageing global population, the increasing prevalence of chronic diseases, the growing demand for minimally invasive procedures, and the ongoing shortage of healthcare professionals. The integration of AI and machine learning capabilities is expected to expand the range of applications where robotics can provide value, whilst cost reductions and improved ease of use will enable adoption in smaller healthcare facilities and emerging markets.

The medical robotics sector represents a compelling investment opportunity for those who understand the regulatory environment, technology requirements, and market dynamics. Companies that can successfully develop safe and effective robotic systems while navigating regulatory approval processes are positioned to capture significant value in this growing market. The integration of humanoid capabilities with medical applications represents a particularly promising area for future development, offering the potential for more flexible, adaptable systems that can address a broader range of healthcare needs.

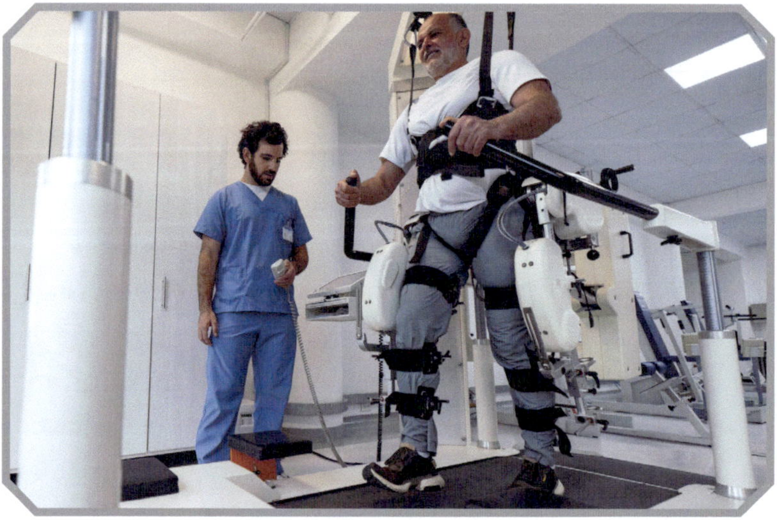

CHAPTER 4:

# Elderly Care Revolution

## Addressing Demographic Challenges

The global ageing population represents one of the most significant demographic challenges of the 21st century, with profound implications for healthcare systems, social services, and economic structures. By 2050, the number of people aged 65 and older is projected to reach 1.6 billion globally, representing 16% of the world's population[15]. This demographic shift is creating unprecedented demand for elderly care services and solutions.

## The Global Ageing Crisis and Robotics Solutions

Traditional elderly care models are becoming unsustainable due to the combination of increasing elderly populations and declining working-age populations in developed countries. Robotics solutions offer the potential to address this crisis by providing consistent, high-quality care while reducing costs and improving outcomes for elderly individuals.

The economic implications of the ageing crisis are staggering. In developed countries, healthcare costs for elderly populations are projected to increase by 50-100% over the next two decades, whilst the working-age population available to support elderly care is shrinking. Traditional approaches to care of the elderly - relying primarily on family caregivers and professional healthcare workers - are becoming increasingly unsustainable as family sizes shrink and healthcare worker shortages intensify.

Elderly care robots can provide 24/7 monitoring and assistance, reduce the physical burden on human caregivers, and enable elderly individuals to maintain independence and dignity whilst living in their own homes. The economic benefits extend beyond direct cost savings to include reduced hospitalisation rates, delayed nursing home placement, and improved quality of life for both elderly individuals and their families.

Japan provides a compelling case study of both the challenges and opportunities presented by population ageing. With over 28% of its population already aged 65 or older, Japan has become a global laboratory for innovations in elderly care robotics. The country's long experiment in automating care has produced valuable insights into the practical applications, social acceptance, and economic benefits of robotic assistance for elderly populations. Japanese companies and research institutions have developed sophisticated companion robots, mobility assistance systems, and monitoring technologies that are now being adopted globally.

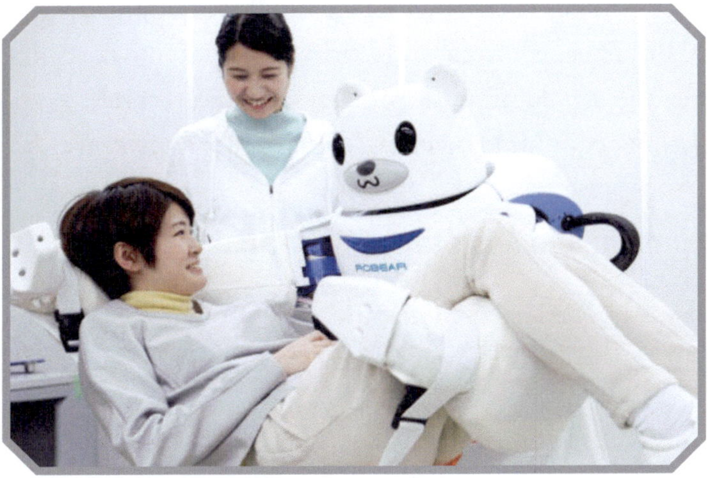

## Demographic Trends and Market Drivers

The global elderly care robotics market is driven by several converging demographic and economic trends. The ratio of working-age adults to elderly individuals is declining rapidly in most developed countries, creating what demographers refer to as the "dependency ratio crisis." In Japan, this ratio has fallen from 8:1 in 1960 to 2.1:1 in 2020 and is projected to reach 1.3:1 by 2050. Similar trends are occurring throughout Europe, North America, and increasingly in other parts of Asia as countries like China and South Korea experience rapid population ageing.

The shortage of professional caregivers is becoming acute in many countries. The United States faces a projected shortage of 355,000 home health aides by 2030, whilst the European Union estimates a shortage of 1.8 million healthcare workers by 2030. These shortages are driving up labour costs and reducing the availability of professional care services, creating compelling economic incentives for robotic alternatives.

## Companion and Assistance Robotics

Elderly care robots demonstrate exceptional utility in mobility assistance, health monitoring, and companionship. These systems can provide 24/7 monitoring, medication reminders, emergency response, and social interaction.

Many companion robots incorporate health monitoring and medication management capabilities, providing gentle reminders about medication schedules, doctor appointments, and health-related activities. These systems can monitor vital signs, track activity levels, and detect changes in behaviour patterns that might indicate health issues.

Advanced AI robots can engage elderly individuals in meaningful conversations, provide reminders about important events or activities, and offer emotional support. These systems can learn individual preferences, remember personal history and relationships, and provide consistent companionship. The therapeutic benefits of companion robots have been documented in numerous studies, showing reduced depression, improved cognitive function, and enhanced overall well-being among users.

Research indicates that elderly individuals often prefer robots that have some human-like characteristics but are clearly identifiable as robots rather than attempting to appear fully human. Cognitive Stimulation and Mental Health Companion robots can provide cognitive stimulation through games, puzzles, memory exercises, and educational activities that help maintain mental acuity and prevent cognitive decline.

## Mobility and Healthcare Support Systems

Mobility assistance represents one of the most practical and immediately beneficial applications of robotics in the care of the elderly. As individuals age, maintaining mobility and independence becomes increasingly challenging due to muscle weakness, balance issues, joint problems, and other age-related physical changes. Robotic mobility assistance systems can provide the support needed to maintain independence whilst reducing the risk of falls and injuries[1].

The development of effective mobility assistance robots requires a sophisticated understanding of human biomechanics, balance control, and the specific challenges faced by elderly individuals with different types of mobility impairments. These systems must be capable of providing appropriate levels of assistance without creating dependency, whilst also ensuring safety and preventing falls or other accidents.

## Cultural Acceptance and Social Integration

The success of elderly care robotics depends not only on technological capabilities but also on cultural acceptance and social integration. Research from Japan and other countries with these advanced care robotics programmes indicates that elderly individuals are generally receptive to robotic assistance, particularly when robots are designed to complement rather than replace human interaction. The key factors influencing acceptance include the robot's appearance and behaviour, the specific tasks it performs, and the way it is introduced and integrated into care routines.

CHAPTER 5:

# Manufacturing and Industrial Applications

## Smart Factory Integration and Industry 4.0

The manufacturing sector represents the largest and most established market for robotics technology, with industrial robotics forming the foundation of modern automated production systems. The global industrial robotics market, valued at $43.8 billion in 2024 with an expected CAGR of 10.5%, continues to evolve rapidly as manufacturers embrace Industry 4.0[16] principles and integrate advanced AI capabilities into their production systems. This transformation extends far beyond traditional robotic automation to encompass intelligent, adaptive systems that can respond to changing production requirements, quality variations, and supply chain disruptions in real time. The integration of humanoid robots into manufacturing environments represents a paradigm shift from fixed, specialised automation systems to flexible, general-purpose platforms that can adapt to human-designed workspaces and utilise standard tools and equipment.

This flexibility addresses one of the fundamental limitations of traditional industrial robotics: the need for extensive customisation and infrastructure modification to accommodate robotic systems. Humanoid robots can work alongside human operators, use existing workstations, and adapt to changing production requirements without requiring major facility modifications.

Manufacturing task success rates for humanoid robots have reached 80-99.5% for selective applications, including pick-and-place, sorting, assembly, and inspection. The achievement of such high success rates reflects advances in computer vision, manipulation control, and AI decision-making systems that enable robots to handle the variability and complexity inherent in real-world manufacturing environments.

## Digital Twin Integration and Predictive Manufacturing

Digital Twins create virtual representations of physical production systems, allowing manufacturers to simulate and optimise robotic operations before implementation, predict maintenance requirements, and identify potential bottlenecks or quality issues before they impact production.

Advanced manufacturing systems now incorporate real-time data from robotic operations into Digital Twin models, creating feedback loops that enable continuous learning and improvement.

## Collaborative Robotics (Cobots) and Human-Machine Interaction

Cobots represent a fundamental shift from traditional industrial robotics, enabling safe human-robot collaboration without the need for safety cages. These systems can work alongside humans, learning from their actions and adapting to changing requirements. Universal Robots has achieved over 100,000 installations of Cobots globally, spanning automotive assembly, electronics manufacturing, food processing, and pharmaceutical production. The success of collaborative robotics reflects both technological advances in safety systems and changing attitudes towards human-robot interaction in manufacturing environments.

## Safety Systems and Human-Robot Collaboration

Advanced safety systems enable Cobots to operate safely in close proximity to human workers. These systems incorporate sophisticated sensor technologies, including vision systems, force sensors, and proximity detectors, that enable robots to detect human presence and adjust their behaviour accordingly. When humans enter the robot's workspace, the system can automatically reduce speed, modify its path, or stop operation entirely to ensure safety.

Modern Cobots can be programmed through demonstration, where human operators physically guide the robot through desired movements, or through simple graphical interfaces that enable rapid reconfiguration for different tasks. This ease of programming reduces the barriers to adoption, enabling manufacturers to deploy collaborative robotics across a broader range of applications.

## Skill Augmentation and Workforce Enhancement Collaborative

Robotics enables a new model of manufacturing work where robots augment human capabilities rather than simply replacing human workers. Robots can handle physically demanding, repetitive, or precision-critical tasks, while human workers focus on problem-solving, quality assessment, and complex assembly operations that require human judgement and dexterity.

The integration of collaborative robotics can enhance worker satisfaction and retention by reducing physical strain, eliminating repetitive tasks, and enabling workers to focus on more engaging, skill-based activities.

## Supply Chain Automation and Logistics

The automation of supply chain and logistics operations represents one of the fastest-growing applications of robotics technology, driven by the explosive growth of e-commerce, increasing consumer expectations for rapid delivery, and the need for greater efficiency and accuracy in warehouse operations. Companies like Symbiotic Inc. have demonstrated the commercial viability of AI-powered warehouse automation systems for retail and e-commerce applications.

Warehouse automation encompasses a range of robotic technologies, including autonomous mobile robots (AMRs), robotic picking systems, automated storage and retrieval systems, and humanoid robots capable of performing a diverse range of logistics tasks. The integration of these technologies creates comprehensive automation platforms that can handle the full range of warehouse operations, from receiving and storage to picking, packing, and shipping.

## Autonomous Mobile Robots and Warehouse Navigation Humanoids

Mobile robots have revolutionised warehouse operations by enabling flexible, scalable automation that can adapt to changing inventory layouts and operational requirements. These systems use advanced navigation technologies to navigate complex warehouse environments whilst avoiding obstacles and coordinating with other robots and human workers.

The deployment of AMRs has enabled warehouse operators to achieve significant improvements in operational efficiency, accuracy, and worker productivity. These systems can operate continuously, handle heavy loads, and navigate efficiently through warehouse environments, reducing the time and physical effort required for material handling operations. The flexibility of AMR systems enables warehouse operators to scale operations up or down based on demand fluctuations without requiring major infrastructure changes.

## Robotic Picking and Manipulation Systems

The development of advanced robotic picking systems addresses one of the most challenging aspects of warehouse automation: the accurate, efficient handling of diverse products with varying shapes, sizes, weights, and packaging.

## Productivity Gains and Economic Benefits

The economic impact of manufacturing robotics extends far beyond direct labour cost savings to encompass improvements in quality, consistency, efficiency, and flexibility that create substantial competitive advantages for manufacturers who successfully implement robotic automation. The compression of business case payback periods from historical 5 to 7 year timelines to current 1 to 3 year periods reflects the improved economic proposition of modern robotic systems[1].

## Quality Improvements and Defect Reduction

Robotic manufacturing systems consistently achieve higher levels of quality and lower defect rates compared to manual operations, particularly for tasks requiring precision, consistency, or operation in challenging environments. The elimination of human variability in critical manufacturing processes can reduce defect rates by 50-90% depending on the application, leading to significant cost savings through reduced rework, warranty claims, and customer returns.

## Operational Efficiency and Throughput Improvements

Robotic systems can operate continuously without breaks, shift changes, or productivity variations associated with human factors, enabling manufacturers to achieve higher levels of operational efficiency and throughput. The ability to operate 24/7 can increase effective production capacity by 40-60% compared to human-operated systems, whilst also reducing the per-unit cost of production through better asset utilisation. Advanced robotic systems can also optimise their own operations through machine learning algorithms that identify opportunities for efficiency improvements, cycle time reductions, and energy savings. These continuous improvement capabilities enable robotic systems to become more efficient over time, providing ongoing benefits that extend well beyond the initial implementation.

## Labour Market Impact and Workforce Transformation

While robots may replace some traditional manufacturing jobs, they also create new opportunities in robot programming, maintenance, supervision, and integration.

The most successful manufacturing robotics implementations focus on augmenting human capabilities rather than simply replacing human workers. This approach leverages the complementary strengths of humans and robots whilst addressing the practical challenges of implementing fully automated systems in complex, variable manufacturing environments.

## Return on Investment and Financial Performance

Comprehensive economic analyses of manufacturing robotics implementations consistently demonstrate positive returns on investment, with typical payback periods ranging from 12 to 36 months depending on the application, labour costs, and operational requirements. The financial benefits encompass direct labour cost savings, quality improvements, efficiency gains, and reduced operational risks that combine to create compelling business cases for robotic automation.

The manufacturing and industrial applications of robotics technology represent a mature, rapidly growing market with demonstrated economic benefits and expanding opportunities. The continued evolution of robotic capabilities, particularly the development of general-purpose humanoid systems, promises to expand these opportunities further while reducing the barriers to adoption for smaller manufacturers and more complex applications.

CHAPTER 6:

# Transportation and Logistics Revolution

The transportation and logistics sector is experiencing a fundamental transformation driven by the convergence of robotics, artificial intelligence, and autonomous systems. This encompasses autonomous vehicles, robotic delivery systems, warehouse automation, and intelligent supply chain management, collectively promising to reshape how goods and people move through the global economy.

## Autonomous Vehicles and Robotic Delivery Systems

Humanoid robots are increasingly being deployed in transportation and logistics applications where their ability to navigate human-designed environments and manipulate diverse objects provides significant advantages over specialised automation systems. These applications range from airport baggage handling and cargo loading to vehicle maintenance and passenger assistance services. The flexibility of humanoid systems enables deployment across multiple transportation modes without requiring extensive infrastructure modifications or specialised equipment.

## Global Robotaxi Outlook: A Transformative Decade Ahead

The global robotaxi market is poised for exponential growth, transforming urban mobility and challenging traditional automotive paradigms. Currently valued at $1.95 billion (2024), the market is projected to reach $43.9 billion by 2033, This represents a CAGR averaging 75%. Leading players in this nascent but rapidly expanding sector include Waymo (Alphabet Inc.), Baidu (Apollo Go), Pony.ai, WeRide, Zoox (Amazon), Wayve (UK), and Tesla. Key operational markets are predominantly in the United States and China, serving as crucial testing grounds and early adoption hubs.

The prevailing ownership model is fleet-based, where companies deploy and manage their own dedicated robotaxi vehicles, offering transportation as a service. Tesla, however, is exploring a hybrid model allowing individual owners to integrate their vehicles into a robotaxi network. Waymo is also considering personal ownership options. Subscription services are also emerging as a potential revenue stream.

The rise of robotaxis presents a significant disruption for conventional car manufacturers. The shift in buying power from individual consumers to large fleet operators will necessitate a reorientation of sales and manufacturing strategies. Increased competition from tech-centric robotaxi companies, a potential decline in private car ownership, and the need for new partnerships or in-house robotaxi divisions will redefine the automotive landscape.

The likely winners will be those who can scale their autonomous technology safely and efficiently, adapt to service-oriented business models, and navigate complex regulatory environments. The technological shift promises a future with reduced traffic congestion, lower transportation costs, and enhanced accessibility, fundamentally altering how people move within cities.

## Autonomous Commercial Vehicles and Freight Transportation

Autonomous trucking systems have achieved significant technological milestones, with companies like Waymo, TuSimple, and Aurora demonstrating successful long-haul operations on public highways. These systems incorporate sophisticated sensor arrays that enable real-time navigation and obstacle avoidance in complex traffic environments. The integration of AI decision-making systems enables these vehicles to handle unexpected situations, weather conditions, and traffic patterns that would challenge traditional automation approaches.

The economic benefits of autonomous commercial vehicles extend beyond labour cost savings to encompass improved fuel efficiency, reduced maintenance costs, enhanced safety records, and optimised routing and scheduling. Autonomous vehicles can operate continuously without mandatory rest periods, potentially increasing asset utilisation by 30-40% compared to human-operated vehicles.

## Last-Mile Delivery and Urban Logistics

Last-mile delivery typically accounts for 40-50% of total shipping costs in e-commerce delivery operations, whilst also being the most carbon-intensive and labour-dependent aspect of the supply chain. Robotic delivery systems offer transformative solutions to these challenges through autonomous ground vehicles, delivery drones, and humanoid robots capable of navigating urban environments and completing complex delivery tasks.

Companies like Serve Robotics Inc., with operations across multiple cities, have demonstrated the commercial viability of autonomous ground delivery vehicles for food delivery, retail goods, and pharmaceutical products. These systems navigate urban environments using advanced computer vision and AI navigation systems, whilst providing secure, contactless delivery services.

## Warehouse Automation and Supply Chain Intelligence

Warehouse automation has become one of the most successful applications of robotics, with companies like Amazon leading the deployment of robotic systems for order fulfilment. These systems can handle millions of items, optimise storage and retrieval, and adapt to changing demand patterns.

This sector has experienced explosive growth, driven by e-commerce expansion, labour shortages, and the need for improved efficiency and accuracy in order fulfilment operations. The global warehouse automation market is projected to reach $69.77 billion by 2025, growing at a CAGR of 14.0%[1] between 2024-2030.

## Mobile Robotic Picking and Manipulation Systems

As mentioned in the previous chapter, Autonomous Mobile Robots and Warehouse Navigation, Autonomous Robile Robots (AMRs) have revolutionised warehouse operations by enabling flexible, scalable automation that can adapt to changing facility layouts and operational requirements. Unlike traditional Automated Guided Vehicles (AGVs) that require fixed infrastructure such as magnetic strips or wire guidance systems, modern AMR systems incorporate sophisticated coordination algorithms that enable multiple robots to work together efficiently, avoiding collisions whilst optimising overall system throughput. These systems can communicate with warehouse management systems to receive real-time task assignments, coordinate with human workers, and provide detailed performance analytics that enable continuous operational optimisation. Robots are now also being used in construction – Monumental, a Dutch company, is making bricklaying robots already[17].

## Maritime Robotics and Autonomous Shipping

The global maritime robotics market is projected to reach \$10 billion by 2026[18], driven by the need for improved safety, environmental compliance, and operational efficiency in maritime operations.

Autonomous shipping represents one of the most ambitious applications of maritime robotics, with several companies developing fully autonomous cargo vessels that can navigate complex maritime environments without human crews.

Port automation has achieved significant commercial success, with automated container terminals demonstrating substantial improvements in operational efficiency, safety, and environmental performance compared to traditional manual operations. Robotic container handling systems can operate continuously, achieve higher precision in container placement, and reduce the safety risks associated with heavy machinery operations in port environments.

Humanoid robots are finding applications in maritime operations where their ability to navigate ship environments and perform complex maintenance tasks provides advantages over specialised systems. These applications include ship inspection and maintenance, cargo handling in confined spaces, and emergency response operations where human-like dexterity and mobility are essential.

## Aviation Robotics and Airport Operations

The aviation sector is integrating robotic systems across multiple operational areas, including aircraft maintenance, ground handling, passenger services, and security operations. The global aviation robotics market is projected to reach $3.9 billion by 2026 driven by the need for enhanced safety, increased efficiency, and reduced costs in aviation operations.

Aircraft maintenance represents a particularly promising application for robotics technology, as these operations require high precision, detailed inspection capabilities, and the ability to work in confined spaces and challenging environments. Robotic inspection systems can perform detailed visual inspections, non-destructive testing, and routine maintenance tasks with greater consistency and accuracy than human technicians, whilst also reducing safety risks and operational costs.

Ground handling operations at airports are increasingly being automated through robotic systems that can handle baggage, cargo, and aircraft servicing.

Humanoid robots are being deployed in airport environments to assist passengers, support security screening, and perform facility maintenance operations.

## Economic Impact and Market Transformation

The economic impact of transportation and logistics robotics extends far beyond direct cost savings to encompass fundamental changes in supply chain structure, urban planning, employment patterns, and global trade flows. The transformation of transportation and logistics through robotics technology represents one of the largest economic opportunities of the 21st century, with implications for virtually every sector of the global economy.

Advanced robotics systems enable supply chain operators to achieve higher levels of efficiency, accuracy, and responsiveness whilst reducing operational costs and improving service quality.

## Urban Planning and Infrastructure Development

The deployment of autonomous vehicles and robotic delivery systems has significant implications for urban planning and infrastructure development. Autonomous vehicles can operate more efficiently than human-driven vehicles, potentially reducing traffic congestion, parking requirements, and infrastructure maintenance costs whilst improving safety and environmental performance.

Robotic delivery systems enable new approaches to urban logistics that can reduce delivery vehicle traffic, improve air quality, and enhance the efficiency of goods movement in dense urban environments. These systems can operate during off-peak hours, utilise alternative transportation routes, and provide more flexible delivery options that improve customer service whilst reducing environmental impact.

## Investment Opportunities and Market Dynamics

The transportation and logistics robotics sector presents compelling investment opportunities across multiple market segments. Key investment considerations include technology maturity, regulatory approval requirements, market adoption rates, and competitive positioning. Developed markets with high labour costs and advanced infrastructure tend to adopt robotic solutions more rapidly, whilst emerging markets may benefit from leapfrogging older technologies and implementing advanced robotic systems directly.

Companies and investors who understand the technological capabilities, market dynamics, and implementation challenges are positioned to benefit from what may be one of the largest technological disruptions in modern economic history.

CHAPTER 7:

# Agricultural Transformation

## Precision Farming and Robotic Agriculture

Agricultural robotics is transforming farming through precision agriculture, autonomous farming systems, and intelligent crop management, and it promises to address some of the most pressing challenges facing global food production: increasing demand from a growing population, climate change impacts, labour shortages, and the need for sustainable farming practices.

The global agricultural robotics market is projected to reach $15.8 billion in 2024, growing at an expected CAGR of 23.3% between 2025 and 2032[22]. This explosive growth reflects the urgent need for technological solutions to agricultural challenges, combined with rapid advances in robotics capabilities that are making automated farming systems commercially viable across a broad range of applications and farm sizes.

Agricultural robots can operate in harsh environmental conditions, work continuously without fatigue, and perform detailed, plant-level interventions that were previously impossible with traditional mechanised farming equipment.

## Precision Agriculture and Autonomous Farming Systems

Autonomous farming systems can perform planting, cultivation, harvesting, and other agricultural tasks with precision that often exceeds human capabilities. These systems use advanced sensors, GPS guidance, and AI to optimise operations and reduce waste.

## AI in Precision Fermentation and Cellular Agriculture

Precision fermentation uses microorganisms to produce specific organic compounds such as proteins and fats through controlled fermentation. Cellular agriculture, by contrast, grows animal cells in laboratory environments to create meat and dairy alternatives, aiming for sustainable food production without traditional farming.

These processes increasingly rely on AI to optimise yield while reducing time, energy, and resource use. AI fine-tunes variables such as vessel dimensions, material composition, temperature, oxygen saturation, and especially the growth medium - which may contain 20 to 200 components. It also enhances the brewing process by optimising timing, sequencing, and delivery conditions, enabling efficient, scalable clean food production.

## Crop Monitoring and Data Collection Systems

Modern agricultural robots serve as sophisticated data collection platforms that can monitor crop health, soil conditions, weather patterns, and pest populations with unprecedented precision and frequency.

Autonomous drone systems have become particularly valuable for large-scale crop monitoring, capable of surveying hundreds of acres per day whilst collecting detailed imagery and sensor data that provides comprehensive insights into crop health and field conditions.

## Planting and Seeding Automation

Robotic planting systems provide precision placement, variable seeding rates, and real-time adaptation to soil conditions, which can enhance germination rates and crop yields while minimising seed waste and input costs. These systems can analyse soil conditions, moisture levels, and topography to optimise seed placement depth, spacing, and timing for maximum germination success.

Advanced planting robots can handle multiple crop types and varieties within the same field, enabling precision agriculture approaches such as companion planting, cover crop integration, and variable seeding rates based on soil quality and expected yield potential.

## Harvesting and Post-Harvest Processing

The development of sophisticated harvesting robots addresses multiple challenges: acute seasonal labour shortages, the physical demands of harvest work, the need for precise timing to optimise crop quality, and the requirement for gentle handling to minimise damage to delicate crops.

Robotic harvesting systems have achieved commercial viability across multiple crop types, from large-scale grain harvesting to delicate fruit and vegetable picking operations.

## Fruit and Vegetable Harvesting Systems

Robotic systems for fruit and vegetable harvesting address some of the most challenging applications in agricultural automation, requiring sophisticated computer vision to identify ripe produce, delicate manipulation to avoid damage, and adaptive behaviour to handle the variability inherent in natural growing environments.

The economic benefits of robotic harvesting extend beyond labour cost savings to encompass improved crop quality, reduced waste, and optimised harvest timing, which can significantly increase the market value of harvested crops. Robotic systems can operate continuously during optimal harvest windows, ensuring that crops are harvested at peak quality rather than being constrained by labour availability or working hour limitations.

## Grain and Row Crop Automation

Large-scale grain and row crop harvesting has achieved high levels of automation through advanced combine harvesters and robotic systems that can operate autonomously across vast agricultural areas. Autonomous harvesting systems can optimise their operations based on crop conditions, weather patterns, and market requirements, adjusting harvesting parameters to maximise yield quality and minimise losses.

## Livestock Management and Animal Husbandry

The integration of robotics into livestock management represents a significant opportunity to improve animal welfare, operational efficiency, and production quality whilst addressing labour shortages and the physical demands of animal husbandry operations. Robotic systems can provide continuous monitoring, automated feeding and milking, health assessment, and facility management.

## Automated Milking and Feeding Systems

Robotic milking systems have achieved widespread commercial adoption in dairy operations, demonstrating significant benefits in milk production, animal health, and operational efficiency. These systems enable cows to be milked on demand rather than on fixed schedules, improving animal comfort and milk production whilst reducing labour requirements and operational costs.

Automated feeding systems can provide individualised nutrition management based on each animal's production stage, health status, and nutritional requirements. These systems can adjust feed formulations, portion sizes, and feeding schedules to optimise animal health and production whilst reducing feed waste and labour requirements.

## Health Monitoring and Veterinary Applications

Robotic systems offer continuous health monitoring capabilities that can detect illnesses, injuries, or behavioural changes in livestock before they become serious problems.

The early detection capabilities of robotic monitoring systems enable prompt veterinary intervention that can prevent disease outbreaks, reduce treatment costs, and improve animal welfare outcomes.

## Environmental Impact and Sustainable Agriculture

Agricultural robotics enables more sustainable farming practices through precise application of inputs, reduced chemical usage, and optimised resource utilisation. These systems can help address global food security challenges while reducing environmental impact.

These robotic systems can analyse soil conditions, crop health, and environmental factors to determine optimal application rates and timing for different areas within a field, reducing waste and environmental impact. Variable-rate application technologies enable farmers to apply inputs only where and when they are needed, reducing overall chemical usage by 20-40% while maintaining crop protection and nutrition levels. This precision approach reduces environmental contamination, lowers input costs, and improves the sustainability of agricultural operations.

Robotic systems can also implement integrated pest management strategies that combine biological controls and targeted chemical applications to reduce reliance on broad-spectrum pesticides. These approaches improve environmental outcomes whilst often providing better pest control results than traditional chemical-intensive methods.

## Soil Health and Conservation Practices

The data collection capabilities of agricultural robots enable detailed monitoring of soil health indicators, carbon sequestration, and ecosystem services that support sustainable agriculture certification and carbon credit programmes. This data can demonstrate the environmental benefits of robotic agriculture whilst providing additional revenue streams for farmers.

## Economic Impact and Market Opportunities

The economic impact of agricultural robotics extends beyond individual farm operations to encompass food security, rural economic development, and global agricultural trade patterns.

The labour savings from robotic systems are particularly significant in regions experiencing acute agricultural labour shortages, where the availability of robotic alternatives can determine the viability of agricultural operations.

Studies indicate that precision agriculture approaches enabled by robotic systems can increase net farm income by 15-25% through improved yields, reduced input costs, and enhanced crop quality.

The data collection and analysis capabilities of robotic systems can improve long-term farm sustainability and profitability while reducing the risks associated with weather variability, market fluctuations, and production challenges.

## Supply Chain

Robotic systems can provide traceability and quality assurance throughout the production process, enabling standardised, high-quality products and improving consumer confidence in food safety.

## Investment Landscape and Market Dynamics

The agricultural robotics sector presents compelling investment opportunities across multiple market segments, from established equipment manufacturers to innovative startups developing specialised solutions.

Key investment considerations include technology maturity, regulatory approval requirements, market adoption rates, and the ability to demonstrate clear economic benefits to farmers.

CHAPTER 8:

# Security and Defence Applications

## Military Robotics and Autonomous Weapons Systems

The defence and security sector represents one of the most strategically significant and rapidly evolving applications of robotics technology, with implications that extend far beyond military operations to encompass national security, geopolitical stability, and the future of warfare itself. Military organisations worldwide are recognising that robotic systems offer unprecedented capabilities for force multiplication, risk reduction, and operational effectiveness that could fundamentally alter the balance of military power and the nature of armed conflict.

The global military robotics market is projected to reach around $30 billion by the end of 2025. This growth is driven by increasing defence budgets, the need to reduce casualties in military operations, and the recognition that autonomous systems provide significant tactical and strategic advantages in modern warfare scenarios.

## Border Security and Surveillance Systems

The application of robotics technology to border security and surveillance is a critical component of national security infrastructure, addressing challenges related to illegal immigration, smuggling, terrorism, and territorial integrity. Robotic surveillance systems provide continuous monitoring capabilities, enhanced detection accuracy, and rapid response coordination, which significantly improve security effectiveness while reducing the personnel requirements and operational costs associated with traditional border security approaches.

## Unmanned Ground Vehicles and Combat Robotics

Ground-based military robots have evolved from simple remote-controlled devices to sophisticated autonomous systems capable of independent navigation, target identification, and engagement in complex battlefield environments.

Modern combat robots can perform a variety of mission types, including reconnaissance, surveillance, explosive ordnance disposal, logistics support, and direct combat operations. The versatility of these systems enables military forces to adapt their capabilities to different mission requirements, reducing the need to deploy specialised equipment or expose human personnel to hazardous situations.

Humanoid military robots represent a particularly significant development, as their human-like form factor enables them to utilise existing military equipment, navigate human-designed environments, and perform complex manipulation tasks that wheeled or tracked robots cannot accomplish. These systems can operate standard military vehicles, use conventional weapons systems, and perform maintenance and logistics tasks in environments designed for human soldiers, as well as those that would be lethal to human soldiers, such as areas with chemical, biological, or radiological contamination.

## Aerial Drones and Unmanned Combat Air Vehicles Military

Aviation has been transformed by the development of unmanned aerial vehicles (UAVs) that range from small tactical drones to large strategic platforms capable of intercontinental operations. These systems have proven their effectiveness in multiple conflict scenarios, demonstrating capabilities for intelligence gathering, surveillance, precision strikes, and air superiority missions that complement and, in some cases, exceed the capabilities of manned aircraft at a fraction of the cost.

## Autonomous Patrol and Monitoring Systems

Robotic patrol systems offer continuous surveillance of borders, critical infrastructure, and sensitive sites, overcoming the limitations of human guards. Equipped with advanced sensors - such as thermal imaging, radar, acoustic detectors, and chemical sensors - these systems can detect threats across multiple modalities. Autonomous ground vehicles and aerial drones efficiently cover large, remote areas in diverse conditions, while also providing detailed documentation of security incidents for investigation and legal purposes.

## Perimeter Security and Access Control

Robotic systems provide enhanced perimeter security capabilities for military installations, government facilities, and critical infrastructure sites. These systems can monitor access points, verify credentials, detect unauthorised intrusions, and coordinate response activities with human security personnel and law enforcement agencies.

Advanced access control robots can perform identity verification using biometric systems, document analysis, and behavioural assessment technologies which provide higher security levels than traditional checkpoint procedures. These systems can also detect concealed weapons, explosives, and other prohibited items using sophisticated scanning technologies.

## Cybersecurity and Information Warfare

As robotic systems become more connected and autonomous, cybersecurity has emerged as a critical concern. These systems are not only vulnerable to cyberattacks but could also be exploited as platforms for launching attacks on other infrastructure. Ensuring their security is essential for both operational integrity and broader system protection.

Robotic cybersecurity spans multiple layers, including hardware protection, software integrity, secure communication protocols, and operational safeguards. Advanced systems now incorporate encrypted communications, secure boot processes, intrusion detection, and autonomous monitoring tools capable of identifying and responding to threats in real time.

## Information Warfare and Robotic Counter Measures

Robotic systems are playing increasingly important roles in information warfare and cyber defence operations, providing capabilities for network monitoring, threat detection, and automated responses to cyberattacks. These systems can analyse vast amounts of network traffic, identify suspicious activities, and coordinate defensive measures faster and more effectively than human operators.

The use of robotic systems in offensive cyber operations raises significant ethical and legal questions about autonomous weapons systems and the rules of engagement in cyber warfare.

## Strategic Implications and Geopolitical Considerations

The proliferation of military robotics technology has profound implications for international security, military balance, and geopolitical stability. The development and deployment of autonomous weapons systems raises fundamental questions about the nature of warfare, the role of human decision-making in lethal force decisions, and the potential for arms races that could destabilise regional and global security.

## Arms Race Dynamics and Military Balance

The rapid advancement of military robotics capabilities is creating new forms of arms race dynamics as nations compete to develop and deploy increasingly sophisticated autonomous weapons systems. The potential for robotic systems to provide decisive military advantages creates incentives for accelerated development and deployment that may outpace the development of appropriate regulatory frameworks and international agreements.

## International Law and Regulatory Frameworks

The development of autonomous weapons systems raises significant questions about compliance with international humanitarian law, the laws of armed conflict, and existing arms control agreements. The ability of robotic systems to make independent decisions about the use of lethal force challenges traditional concepts of human responsibility and accountability in warfare.

International efforts to develop regulatory frameworks for autonomous weapons systems face significant challenges due to the rapid pace of technological development, differing national security interests, and the dual-use nature of many robotic technologies.

## Economic Impact and Defence Industrial Base

The military robotics sector represents a significant component of the global defence industry, with implications for defence spending, industrial capacity, and technological competitiveness.

Military robotics represents an increasing share of defence budgets as nations recognise the strategic importance of autonomous systems and seek to maintain technological superiority in this critical area. The US Department of Defense has so far allocated over $3.7 billion for unmanned systems research and development in fiscal year 2025, reflecting the priority placed on robotic capabilities.

## Innovation and Technology Transfer

Military robotics research and development drives innovation across multiple technology sectors, including artificial intelligence, advanced materials, sensor technologies, and manufacturing processes. The demanding requirements of military applications often push technological capabilities beyond what civilian applications require, leading to advances that subsequently benefit civilian markets.

The strategic importance of military robotics ensures that this sector will continue to receive significant investment and attention from governments, defence contractors, and technology companies worldwide. The successful development and deployment of military robotic systems require careful consideration of technological capabilities, operational requirements, ethical implications, and strategic consequences that extend far beyond the immediate military applications to encompass broader questions of security, stability, and the future of warfare itself.

CHAPTER 9:

# Domestic Help and Service Robotics

## Home Automation and Personal Assistance - The Consumer Revolution

The domestic robotics sector represents one of the most visible and rapidly growing applications of robotics technology, bringing advanced automation capabilities directly into consumers' homes and daily lives. This sector encompasses cleaning robots, personal assistance systems, home security robots, and companion devices that collectively promise to transform household management and enhance quality of life for millions of families worldwide.

The global domestic robotics market is projected to reach $34.1 billion by the end of 2025, growing at an expected CAGR of 9.3% from then to 2030[1]. This growth is driven by increasing consumer acceptance of robotic technology, declining costs of robotic systems, and the development of more sophisticated AI capabilities that enable robots to perform complex household tasks with minimal human supervision.

Consumer robotics represents a unique market segment where user experience, safety, and reliability requirements often exceed those of industrial applications. Domestic robots must operate safely around children, pets, and valuable household items whilst providing intuitive interfaces that enable users without technical expertise to operate and maintain these systems effectively.

## Cleaning and Maintenance Robotics

Robotic vacuum cleaners have achieved widespread consumer adoption, with companies like Beijing Roborock Technology achieving nearly $5 billion in market capitalisation through innovative cleaning robot designs. These systems have evolved from simple automated vacuum cleaners to sophisticated home maintenance platforms that can map household layouts, avoid obstacles, and coordinate with other smart home systems.

Modern cleaning robots incorporate advanced navigation systems, multi-surface cleaning capabilities, and AI-powered scheduling, enabling them to maintain household cleanliness with minimal human intervention. The integration of smart home connectivity enables these systems to coordinate with other household devices, providing detailed cleaning reports and maintenance alerts.

The expansion of cleaning robotics beyond vacuum cleaners to include mopping, window cleaning, and outdoor maintenance systems represents significant growth opportunities as consumers seek comprehensive automated solutions for household maintenance tasks.

## Entertainment and Companion Robotics

The development of entertainment and companion robots addresses growing consumer interest in interactive technology and the need for social engagement, particularly among elderly populations and families with limited social interaction opportunities. These systems combine advanced AI capabilities with engaging personalities and interactive features, providing entertainment, education, and emotional support.

The integration of natural language processing, facial recognition, and emotional intelligence capabilities enables modern companion robots to provide personalised interaction that adapts to individual preferences and needs. These systems can learn family routines, remember personal preferences, and provide customised responses that enhance the user experience over time.

## Educational and Child Development Applications

Educational robots provide interactive learning experiences that can enhance child development and academic achievement through engaging, hands-on activities. These systems can teach programming concepts, STEM subjects, and social skills through play-based learning that adapts to individual learning styles and progress rates.

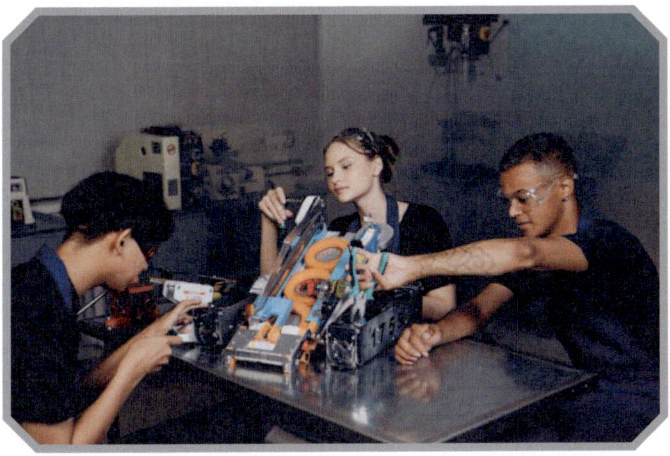

The ability of educational robots to provide consistent, patient instruction makes them particularly valuable for children with learning disabilities or attention challenges who may benefit from alternative teaching approaches. These systems can provide unlimited repetition, customised pacing, and positive reinforcement that support learning success.

## Market Opportunities and Consumer Adoption

The consumer robotics market presents unique opportunities and challenges compared to industrial applications. Consumer price sensitivity, safety requirements, and user experience expectations create distinct design and marketing considerations that necessitate specialised approaches to product development and market positioning.

The successful consumer robotics companies have demonstrated the importance of focusing on specific use cases where robots provide clear value propositions, such as time savings, improved results, or enhanced convenience. The most successful products often begin with simple, well-defined tasks and gradually expand their capabilities as consumer acceptance and technological maturity increase.

The geographic distribution of consumer robotics adoption reflects income levels, technology acceptance, and cultural attitudes towards automation. Developed markets with high labour costs and technology-savvy consumers tend to adopt domestic robotics more rapidly, whilst emerging markets may benefit from leapfrogging traditional appliances and implementing robotic solutions directly.

In short, investors who don't look at robots and associated investments as being key to their long-term portfolio, will be missing out on **one of the greatest industries ever**. All of us will probably own a robot or lease one. All of us will benefit, but not many will benefit from early investment in them. This is our chance.

APPENDIX A:

# Listed Robot Manufacturers

| Ticker | Name | Country | Product | Market cap ($m) | Sector |
|---|---|---|---|---|---|
| ABBN SW | ABB LTD-REG | SWITZER-LAND | ABB YuMi collaborative robot, IRB series industrial robots | $132,030 | Industrial |
| BA/LN | BAE SYSTEMS PLC | BRITAIN | Remote controlled combat vehicles | $82,424 | Defence |
| CNH US | CNH INDUSTRIAL NV | BRITAIN | Agricultural Robots | $13,671 | Agricultural |
| 6954 JP | FANUC CORP | JAPAN | Industrial robots for welding, packaging and assembley | $28,592 | Industrial |
| GD US | GENERAL DYNAMICS CORP | UNITED STATES | Autonomous defense robotics | $87,799 | Defence |
| 7267 JP | HONDA MOTOR CO LTD | JAPAN | Asimo humanoid robot | $59,732 | Industrial |
| HON US | HONEYWELL INTERNATION-AL INC | UNITED STATES | Warehouse robotics, autonomous mobile robots (AMRs) | $134,433 | Industrial |
| 005380 KS | HYUNDAI MOTOR CO (BOSTON DYNAMICS) | SOUTH KOREA | Acquired Boston Dynamics (Atlas,Spot,Stretch) | $31,859 | Industrial |
| ISRG US | INTUITIVE SURGICAL INC | UNITED STATES | Robotic-assisted surgery system | $155,575 | Medical |
| IRBT US | IROBOT CORP | UNITED STATES | Consumer and military robots | $126 | Industrial |

| | | | | | |
|---|---|---|---|---|---|
| 7012 JP | KAWASAKI HEAVY INDUSTRIES | JAPAN | Industrial robots Softbank - Pepper social robot | $10,510 | Industrial |
| KSCP US | KNIGHTSCOPE INC-A | UNITED STATES | Security robots | $60 | Consumer |
| 066570 KS | LG ELECTRONICS INC | SOUTH KOREA | Home service robots | $9,186 | Domestic |
| LMT US | LOCKHEED MARTIN CORP | UNITED STATES | Unmanned ground and aerial robotics systems | $110,487 | Defence |
| MDT US | MEDTRONIC PLC | IRELAND | Surgical robotic systems for minimally invasive operations | $119,841 | Medical |
| 000333 CH | MIDEA GROUP CO LTD-A (KUKA) | CHINA | Acquired KUKA Robotics, producing industrial and humanoid-ready robot | $80,551 | Industrial |
| NOC US | NORTHROP GRUMMAN CORP | UNITED STATES | Unmanned aerial vehicles | $82,264 | Defence |
| 005930 KS | SAMSUNG ELECTRONICS CO LTD | SOUTH KOREA | Service robotics | $335,731 | Domestic/ consumer |
| SIE GY | SIEMENS AG-REG | GERMANY | Automation robotics integrated into factory systems | $217,165 | Industrial |
| SYK US | STRYKER CORP | UNITED STATES | Robotic system for orthopedic surgery | $144,172 | Medical |
| TER US | TERADYNE INC | UNITED STATES | Owns Universal Robots (cobots) and Mobile Industrial Robots (AMRs) | $18,146 | Industrial |
| TSLA US | TESLA INC | UNITED STATES | Optimus humanoid programme | $1,401,144 | Industrial/ consumer |
| 7203 JP | TOYOTA MOTOR CORP | JAPAN | T-HR3 humanoid, WelWalk rehab robots, AI mobility assistants | $315,180 | Industrial |
| 9880 HK | UBTECH ROBOTICS CORP LTD-H | CHINA | Walker humanoid, educational and household robots | $7,150 | Domestic/ consumer |
| 1810 HK | XIAOMI CORP-CLASS B | CHINA | Xiaomi CyberOne humanoid, CyberDog quadruped | $188,802 | Domestic/ consumer |
| 6506 JP | YASKAWA ELECTRIC CORP | JAPAN | Industrial robot series for manufacturing tasks | $5,509 | Industrial |
| ZBH US | ZIMMER BIOMET HOLDINGS INC | UNITED STATES | Robotic surgical assistant for spine and knee surgery | $20,000 | Medical |

APPENDIX B:

# Listed Component Manufacturers

| Ticker | Name | Country | Product | Market cap ($m) | Sector |
|---|---|---|---|---|---|
| 6324 JP | HARMONIC DRIVE SYSTEMS INC | JAPAN | Harmonic reducers | $1,705 | Precision reducers |
| 2049 TT | HIWIN TECHNOLOGIES CORP | TAIWAN | Linear guides, ball screws | $2,530 | Linear guides and ball screw |
| 6680 HK | JL MAG RARE-EARTH CO LTD -H | CHINA | High performance magnets | $6,342 | Rare-earth magnets |
| 179 HK | JOHNSON ELECTRIC HOLDINGS | HONG KONG | Miniature motors | $4,337 | Miniature motors |
| 6861 JP | KEYENCE CORP | JAPAN | Vision & force sensors | $93,721 | Vision systems & sensors |
| 688017 CH | LEADER HARMONIOUS DRIVE SY-A | CHINA | Harmonic drives | $4,649 | Harmonic drives |
| 6268 JP | NABTESCO CORP | JAPAN | RV reducers, actuators | $2,635 | RV reducers |
| 6594 JP | NIDEC CORP | JAPAN | Servo motors, actuators | $21,660 | Motors & actuators |
| 6471 JP | NSK LTD | JAPAN | Bearings, ball screws | $2,730 | Ball screws & bearings |
| OUST US | OUSTER INC | UNITED STATES | Lidar sensors | $1,675 | LiDAR sensors |

| | | | | | |
|---|---|---|---|---|---|
| RRX US | REGAL REXNORD CORP | UNITED STATES | Motors, reducers, actuators | $9,363 | Motors & actuators |
| SHA0 GY | SCHAEFFLER AG | GERMANY | Bearings and mechanical parts | $6,357 | Bearings & mechanical part |
| 003021 CH | SHENZHEN ZHAOWEI MACHINERY-A | CHINA | Micro gear motors | $4,810 | Micor gear motors |
| SKFB SS | SKF AB-B SHARES | SWEDEN | Bearings, actuation components | $11,560 | Bearings |
| TDY US | TELEDYNE TECHNOLO-GIES INC | UNITED STATES | High-resolution cameras | $26,101 | Vision and imaging sensors |

APPENDIX C:

# Private Companies

| Ticker | Name | Country |
|---|---|---|
| 1X TECHNOLOGIES | NORWAY | EVE/NEO - Domestic, services,security |
| AGIBOT | CHINA | Industrial humanoid robots |
| AGILITY ROBOTICS | US | Digit - Multi purpose |
| ANYBOTICS | SWITZERLAND | ANYmal - oil, gas, & utilities inspection |
| APPTRONIK | US | Apollo - Warehouse/factory |
| BLUE RIVER TECHNOLOGY | US | Agricultural robots |
| DILIGENT ROBOTICS | US | Moxi - Medical assistant |
| ENGINEERED ARTS | UK | Ameca - research, exhibitions |
| FIGURE AI | US | Helix - multi purpose |
| FOURIER INTELLIGENCE | CHINA | Medical rehab, exoskeletons |
| GALBOT | CHINA | Warehouse and service robots |
| GITAI | JAPAN/US | Assembly |
| KEPIER | CHINA | Humanoid robots |
| NURO | US | Autonomous delivery |
| PAL ROBOTICS | SPAIN | Retail, research and assistance |
| RELAY ROBOTICS | US | Delivery robots for hospitals and hotels |
| ROBOCORE TECH | CHINA | Temi - multi purpose |

| | | |
|---|---|---|
| SANCTUARY AI | CANADA | Phoenix - general purpose, retail |
| STARSHIP TECHNOLOGIES | US/Estonia | Autonomous delivery |
| UNITREE ROBOTICS | CHINA | Logistics, research |
| UNIVERSAL ROBOTS | US | Collaborative robot Festo - automation technology |

## Venture Capital Funds

Andreessen Horowitz, Khosla Ventures, GV (Google Ventures), Intel Capital

APPENDIX D:

# Key Opinion Leaders in Robotics and Humanoids

## Academic Leaders and Researchers

**Prof. Dieter Fox** - University of Washington German roboticist and Professor in Computer Science & Engineering. Co-author of "Probabilistic Robotics" with Wolfram Burgard and Sebastian Thrun. Recently joined NVIDIA to start a Robotics Research Lab in Seattle. Fellow of AAAI and IEEE, former editor of IEEE Transactions on Robotics.

**Davide Scaramuzza** - University of Zurich Founder and director of the Robotics and Perception Group. Professor of Robotics and Perception at Zurich University. Research focus on autonomous drone navigation, computer vision, and neuromorphic cameras. Co-founded Zurich-Eye for visual-inertial navigation solutions.

**Prof. John J. Leonard** - MIT Member of MIT Computer Science and Artificial Intelligence Laboratory (CSAIL). Renowned SLAM (Simultaneous Localisation and Mapping) researcher. Team leader for MIT's 2007 DARPA Urban Challenge team (4th place). Currently VP of Autonomous Driving Research at Toyota Research Institute.

**Prof. Juan D. Tardós** - University of Zaragoza, Spain Professor at Departamento de Informática e Ingeniería de Sistemas. Research group pioneering robot vision, SLAM, perception, and mobile robotics. Member of I3A Robotics, Perception and Real-Time Group.

**Jitendra Malik** - UC Berkeley Professor of Computer Science, EECS Department. Research in computer vision, computer graphics, and biological-image analysis. Multiple award winner including Longuet-Higgins Award (2007, 2008) and PAMI Distinguished Researcher Award in Computer Vision (2013).

**Prof. Jun Ho Oh** - KAIST, South Korea Head of KAIST Humanoid Research Center and co-founder of Rainbow Robotics Co. Winner of 2015 DARPA Robotics Challenge with "Hubo" humanoid robot ($2M prize). Pioneer credited with starting robotics research in Korea.

**Prof. Peter Corke** - Queensland University of Technology, Australia Director of Australian Research Council Center of

Excellence in Robotic Vision. Former editor-in-chief of Robotics & Automation magazine (2009-2013).

**Prof. Masayuki Inaba** - University of Tokyo, Japan Professor of Information Science and Technology. Leads Tokyo's famous "JSK lab." Students created Google's SCHAFT robotics company. Multiple awards from Robotics Society of Japan.

## Industry Thought Leaders

### Top 10 Industry Influencers

- Jair Pinto Ribeiro - Volvo Group
- Prof. Marek Kowalkiewicz - QUT (Queensland University of Technology)
- Marsha Collier - WS Radio
- Luca Collina MBA - TRANSFORAGE TCA Ltd
- Tom Raftery - Independent Technology Analyst
- Scott Luton - Supply Chain Now
- Nicolas Babin - Babin Business Consulting
- Lina Al Atabani - Independent Consultant
- Professor Andy Pardoe - Informed AI
- Tony Rhem - A.J. Rhem & Associates, Inc.

## Notable Industry Leaders

- Elmustafa Sayed Ali - MASAD
- Olivier Gomez (OG) - IAC – Intelligent Automation Company
- Nikhil Malhotra – Tech Mahindra Ltd
- Vladimer Botsvadze - Technology Consultant
- Dinis GUARDA - ztudium
- Hessie Jones - MyData Canada
- Adi Gaskell - The Horizons Tracker
- Michael Gale - EDB
- Michael Krigsman - CXOTalk
- Kevin Delaney - Cisco
- Rafael Domene - Independent
- Logan Harper, MS - Boston Children's Hospital
- Christophe Foulon - CPF Coaching
- Danilo McGarry - Technology Consultant
- Andrey Golub - ELSE Corp, a Digital Retail LAB

## Historical Pioneers and Visionaries Foundational Figures

- Isaac Asimov - Science fiction author who created the Three Laws of Robotics
- Joseph Engelberger - "Father of Robotics," created first industrial robot
- George Devol - Inventor of the first digitally operated programmable robotic arm
- Marvin Minsky - AI pioneer and co-founder of MIT's AI laboratory

## Contemporary Visionaries

- Elon Musk - Tesla (Optimus humanoid robot program)
- Marc Raibert - Boston Dynamics founder
- Rodney Brooks - Former MIT robotics professor, iRobot co-founder
- Cynthia Breazeal - MIT, pioneer in social robotics and human-robot interaction

## Corporate Leaders Major Robotics Companies CEOs and Founders

- Melonee Wise - Former CEO of Unbounded Robotics
- Steve Cousins - CEO of Savioke, formerly of Willow Garage
- Brian Gerkey - CEO of Open-Source Robotics Foundation
- Colin Angle - CEO and co-founder of iRobot
- Ryan Gariepy - CTO and co-founder of Clearpath Robotics

## Robotics Technology Platform Leaders

- Jensen Huang - NVIDIA CEO (Robotics AI platforms)
- Satya Nadella - Microsoft CEO (Azure AI for Robotics)
- Sundar Pichai - Google CEO (AI and robotics research)
- Andy Jassy - Amazon CEO (Warehouse robotics and AI)

## Emerging Leaders in Humanoid Robotics Chinese Humanoid Robotics Leaders

- Wang Xingxing - Unitree Robotics founder
- Xiong Youjun - UBTech founder and chairman
- Chen Jianyu - Galbot founder
- Hu Huazhi – Founder of EHang

## Global Humanoid Robotics Innovators Agility Robotics

- Damion Shelton - Co-founder of Agility Robotics
- Brett Adcock – CEO Figure AI
- Bernt Øivind Børnich (Founder) @1x
- Geordie Rose (co-founder) Sanctuary AI

*Note: This list represents key opinion leaders who have significantly influenced the development, commercialisation, and strategic direction of robotics and humanoid technologies as of 2024-2025. Their contributions span academic research, industry innovation, corporate leadership, and technological advancement.*

Appendix E:

# Research Sources and Citation
# Primary Sources


1. Citigroup.com. (2024). *The Rise of AI Robots*. [online] Available at: https://www.citigroup.com/global/insights/the-rise-of-ai-robots.

2. International Monetary Fund (2025). *GDP, Current Prices*. [online] International Monetary Fund. Available at: https://www.imf.org/external/datamapper/NGDPD@WEO/OEMDC/ADVEC/WEOWORLD.

3. Ia`FR International Federation of Robotics (2024). *Record of 4 Million Robots in Factories Worldwide*. [online] IFR International Federation of Robotics. Available at: https://ifr.org/ifr-press-releases/news/record-of-4-million-robots-working-in-factories-worldwide.

4. Exportsemi.com. (2025). *China's Humanoid Robot Industry in 2025: Booming Orders Amid a Growing Talent Crisis.* [online] Available at: https://www.exportsemi.com/company-post/robot-orders-to-surge-in-2025-wang-xingxing-reveals-major-talent-gap-at-unitree/

5. Research and Markets (2025). *Humanoid Robot Research Report 2024: Market to Grow by Over $110 Billion - Opportunities and Strategies to 2033*. [online] GlobeNewswire News Room. Available at: https://www.globenewswire.com/news-release/2025/04/04/3055807/28124/en/Humanoid-Robot-Research-Report-2024-Market-to-Grow-by-Over-110-Billion-Opportunities-and-Strategies-to-2033.html.

6. Exotec. (2025). *Complete Guide to Pick-and-Place Automation.* [online] Available at: https://www.exotec.com/insights/complete-guide-to-pick-and-place-automation/

7. Saeed, A., Kelkar, A., Jansen, C., Tilley, J., Patel, M. and Chui, M. (2025). *Will embodied AI create robotic coworkers?* [online] McKinsey & Company. Available at: https://www.mckinsey.com/industries/industrials-and-electronics/our-insights/will-embodied-ai-create-robotic-coworkers.

8. South China Morning Post. (2025). *South China Morning Post.* [online] Available at: https://www.scmp.com/tech/article/3320679/beijings-e-town-offers-subsidies-robot-buyers-china-embraces-humanoids

9. FELT, C. (2020). *Autonomous Weaponry: Are Killer Robots in Our Future?* [online] The Henry M. Jackson School of International Studies. Available at: https://jsis.washington.edu/news/autonomous-weaponry-are-killer-robots-in-our-future/.

10. Kelkar, A. and Luchtenberg, D. (2025). *The robotics revolution: Scaling beyond the pilot phase*. [online] McKinsey & Company. Available at: https://www.mckinsey.com/capabilities/operations/our-insights/the-robotics-revolution-scaling-beyond-the-pilot-phase .

11. Edwards, D. (2025). *Humanoid robot market to grow to $110 billion in next eight years*. [online] Robotics & Automation News. Available at: https://roboticsandautomationnews.com/2025/04/05/humanoid-robot-market-to-grow-to-110-billion-in-next-eight-years/89595/

12. Jonas, A. and Zhong, S. (2025). *A $5 Trillion Global Market.* Morgan Stanley.

13. Efficiency AI Transformation - AI Process Consultancy. (2025). *Neuromorphic Computing for Robotics Explained, AI Consultants UK*. [online] Available at: https://www.efficiencyai.co.uk/knowledge_card/neuromorphic-computing-for-robotics/

14. Giiresearch.com. (2024). *Medical Robotic Systems Market Size, Share & Trend Analysis Report By Type (Cleanroom Robots, Robotic Prosthetics, Medical Service Robots), Region, and Segment*

*Forecasts, 2024 - 2030*. [online] Available at: https://www.giiresearch.com/report/grvi1552610-medical-robotic-systems-market-size-share-trend.html

15. United Nations (2023). *Leaving No One Behind in an Ageing World - World Social Report*. United Nations

16. McKinsey (2022). *What are Industry 4.0, the Fourth Industrial Revolution, and 4IR?* [online] McKinsey & Company. Available at: https://www.mckinsey.com/featured-insights/mckinsey-explainers/what-are-industry-4-0-the-fourth-industrial-revolution-and-4ir

17. Kelly, E. (2025). *Construction Robots - Here come the builder bots*. Sifted and Leonard.

18. Research and Markets (2022). *Global Autonomous Ships Market Report 2022-2026 - China Aims at Being Leading Autonomous Shipping Country*. [online] GlobeNewswire News Room. Available at: https://www.globenewswire.com/news-release/2022/06/10/2460476/28124/en/Global-Autonomous-Ships-Market-Report-2022-2026-China-Aims-at-Being-Leading-Autonomous-Shipping-Country.html

19. www.imarcgroup.com. (n.d.). *Robotics Market Size, Growth Trends, Report and Forecast 2022-2027*. [online] Available at: https://www.imarcgroup.com/robotics-market.

20. Market.us. (n.d.). *Industrial Robotics Market Size, Share | CAGR of 11.4%*. [online] Available at: https://market.us/report/industrial-robotics-market/.

21. Exactitude Consultancy and Exactitude Consultancy (2025). *Medical Robotics Market Research Report By Product Type (Surgical Robots, Rehabilitation Robots, Exoskeletons, Telepresence Robots), By Application (Orthopedic Surgery, Neurology, Cardiology, Gynecology), By End User (Hospitals, Ambulatory Surgical Centers, Rehabilitation Centers), By Technology (Robotic-Assisted Surgery, Rehabilitation Robotics, Autonomous Robotics), By Distribution Channel (Direct Sales, Distributors, Online Sales) – Forecast to 2034.* [online] Exactitude Consultancy. Available at: https://exactitudeconsultancy.com/reports/73221/medical-robotics-market [Accessed 29 Sep. 2025].

22. Agritechtomorrow.com. (2024). *Agricultural Robotics: Powering the Future of Smart and Sustainable Farming | AgriTechTomorrow*. [online] Available at: https://www.agritechtomorrow.com/story/2025/09/agricultural-robotics-powering-the-future-of-smart-and-sustainable-farming/16956/ [Accessed 29 Sep. 2025].

## Additional Research Resources

- International Federation of Robotics - Global Robot Statistics 2023

- Morgan Stanley Research - Humanoid Robotics Market Projections 2024
- McKinsey Global Institute - Robotics and AI Research 2025
- 2025 World Robot Conference Beijing - Industry Reports
- NVIDIA Robotics Research Publications
- MIT CSAIL Robotics Research Papers
- IEEE Transactions on Robotics
- Morgan Stanley. (2025). *Future State: 1 Billion Humanoids*
- Morgan Stanley. (2024) *Humanoid Helpers of the Future*


## Recent Publications Industry Reports


- Boston Consulting Group - Robotics Market Analysis 2024
- PwC Global Robotics Survey 2025
- Deloitte Technology Trends - Robotics Section 2025
- Accenture Future of Work - Robotics Integration 2024


## Academic Publications


- Nature Robotics - Recent Research Articles
- Science Robotics - Breakthrough Technologies 2024-2025

- Journal of Field Robotics - Commercial Applications
- Robotics and Autonomous Systems


## Technical Advances Company Reports and Filings


- Tesla Annual Report 2024 - Optimus Program
- Boston Dynamics Technical Publications
- Universal Robots Market Data 2025
- Chinese Robotics Companies


## Public Filings 2025 Government and Policy Sources


- US Department of Defence - Robotics Strategy 2024
- European Commission - Robotics Policy Framework
- Chinese Ministry of Industry - Robotics Development Plan
- Japanese Society for the Promotion of Science - Robotics Research


*All sources accessed and verified as of August 2025. This document represents an independent analysis based on publicly available information and industry research.*  *This document contains forward-looking statements and projections*

APPENDIX F:

# Highly Rated Books on Robotics, Humanoids, and Artificial Intelligence

## Essential Reading

1. "Probabilistic Robotics" by Sebastian Thrun, Wolfram Burgard, and Dieter Fox
2. "Introduction to Autonomous Mobile Robots" by Roland Siegwart and Illah Nourbakhsh
3. "Robotics: Modelling, Planning and Control" by Bruno Siciliano and Lorenzo Sciavicco, et al
4. "The Robotics Primer" by Maja Matarić
5. "Rise of the Robots" by Martin Ford AI and Future Perspectives
6. "Superintelligence: Paths, dangers, strategies" by Nick Bostrom

7. "Life 3.0" by Max Tegmark

8. "The Second Machine Age: Work, Progress, and prosperity in a time of brilliant technologies " by Erik Brynjolfsson and Andrew McAfee

9. "Human Compatible: artificial intelligence and the problem of control" by Stuart Russell

10. "The Alignment Problem" by Brian Christian

11. "Fuel-cell-powered robot platforms by J Shepherd online at http://www.batterypowertips.com/fuel-cell-powered-robot-platforms-faq

# MEET JIM MELLON

Jim Mellon is a British entrepreneur, investor, and author, with interests in a number of sectors. He has substantial real estate assets in Germany and the Isle of Man, as well as holdings in private and public companies through the Burnbrae Group, his private investment company.

After leaving Oriel College, Oxford, where he studied Philosophy, Politics and Economics, he worked in Asia and the United States in two fund management companies, GT Management and Thornton Management (Asia) Limited, before founding Regent Pacific Group Limited in 1991, subsequently quoted on the Hong Kong Stock Exchange.

Jim spends most of his time investigating and working on start-up ideas, inspired by his ability to identify technologies and companies that will reshape people's lives, before they start trending. His investment philosophy of focussing on these hyper-innovative technologies and novel companies has led him to be recognised as one of the most successful investors of his generation.

He is the author of several books, mostly written with a view to identifying emerging thematic trends leading to early-stage investment opportunities. Notably, in his book *Wake Up! Survive and Prosper in the Coming Economic Turmoil* (2005), he forecast the global financial crisis of 2008-2009. His book *Juvenescence: Investing in the Age of Longevity* (2017) played a role in bringing ageing research into the mainstream. It led to the formation of the company of which he is co-founder and chairman, Juvenescence Ltd, which develops therapies that will enable people worldwide to live longer, healthier lives.

Jim is deeply passionate about the fight against animal cruelty. He is a pescatarian and dog-lover and has set his eye on the alternative proteins space having come to realise the alarming level of food scarcity in the world and the environmental damage industrial farming has done to our planet. He is the founder and director of Agronomics, a London-listed investment company which is building a portfolio of investments in the most promising companies in the field of clean food.

Mr Mellon is the executive chairman of the board of Manx Financial Group plc, and he is also a non-executive director of Bradda Head Investments and Chairman of Regent Pacific Group.

Jim is a co-founder and trustee of The Longevity Forum and the Longevity Biotechnology Association. He is also on the Board of the Buck Institute for Research on Aging. He is a Fellow of Oriel College, University of Oxford.